FOSS Science Resources

Insects and Plants

Full Option Science System
Developed at
The Lawrence Hall of Science,
University of California, Berkeley
Published and distributed by
Delta Education,
a member of the School Specialty Family

1487699
978-1-62571-291-2
Printing 8 — 4/2019
Standard Printing, Canton, OH

Table of Contents

Animals and Plants in Their Habitats

Look at this **grassland habitat**. Do you see anything **living** here? Let's take a closer look.

Look! A grasshopper sits in the green grass.
Grass **plants** and grasshoppers are living in this
grassland habitat. Grass and grasshoppers get
their **basic needs** met in the grassland habitat.

Land **animals** need food, water, air, space, shelter, and a comfortable temperature. Grasshoppers eat grass for food and water. In the grass, grasshoppers have air and shelter. In summer, the grassland provides a warm habitat for **insects**, like grasshoppers, and other animals, like prairie dogs.

Plants need water, air, sunlight, and **nutrients**. Grass gets water and nutrients from the rich, moist grassland soil. Their roots take up water and nutrients. Air flows around the grass. Grass uses sunlight to make the food it needs to live.

Other parts of the country are different
from the grassland habitat. This is a
desert habitat. What do you see living
here? Let's take a closer look.

Plants and animals are living in the desert, too. How do they get the things they need to live?

Desert plants get water and nutrients from the desert soil. The plants have large root systems to collect water. Some desert plants, like cactuses, have thick spongy stems. Cactus stems store water for the plant to use later.

The Sun shines brightly on the desert most of the time. Air flows easily around the desert plants. Deserts are very hot during the summer. But desert plants can live even in very hot temperatures.

Some desert animals eat other animals for food. Some desert animals eat **seeds** for food. Harvester ants gather seeds. They store the seeds to eat during the year. Ants get water from the seeds they eat.

Ants dig tunnels and chambers. Their underground tunnels give ants a safe space to live. The tunnels provide shelter for the ants so they can **survive** the hot desert summer.

This is a **tundra** habitat in winter.
Some parts of Alaska have this kind
of habitat. What do you see living
here? Let's take a closer look.

Tundra plants survive the winter in a resting stage. When the Sun warms the land, the plants begin to grow again. The plants take up water and nutrients from melting ice and snow. They use sunlight to make their food. Tundra plants can survive cold temperatures.

Caribou roam the open space of the tundra. They eat the short plants. They drink from pools of melted snow. Caribou have thick fur to protect them from the cold temperature. But the fur is not thick enough to protect them from mosquitoes.

Mosquitoes burrow down in tundra plants in the fall. They rest there during the cold winter. When the snow melts in spring, the mosquitoes look for food and mates. The female mosquitoes lay **eggs** on pools of water.

After the eggs hatch, mosquito **larvae** eat tiny bits of food in the water. The mosquito larvae grow quickly for a few weeks. Then they swim to the surface of the water and break out of their old skin. Now they are **adult** mosquitoes and fly to find food.

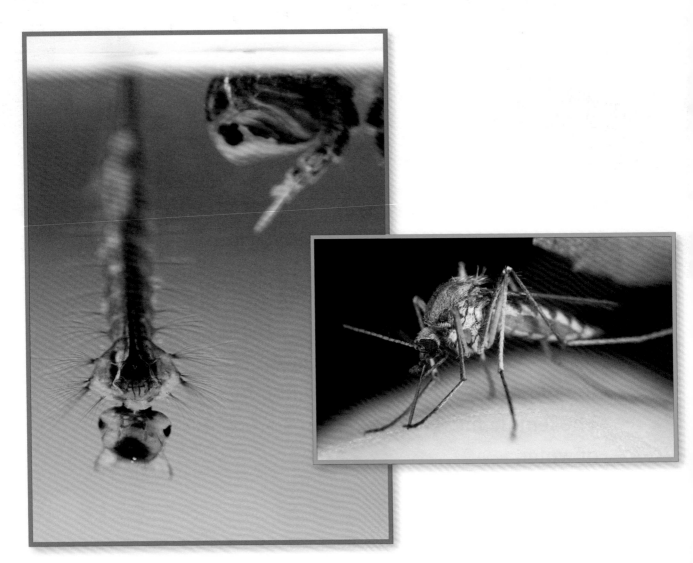

When you look closely at a habitat, you will find animals and plants living there. Living things **thrive** when they have what they need to live.

Flowers and Seeds

These are wild brassica plants. Each
plant grows a lot of **flowers**. But
brassica plants do not grow flowers to
look pretty. The flowers are an important
part of the plant's **life cycle**.

Soon, the flowers fade and dry up.
Something new appears right where
each flower once grew. It looks like a
little green bean. It is a seedpod.

Weeks later, the seedpods are big and dry.
There are about six seeds inside each seedpod.
What do you think will happen if someone
plants the new seeds?

Brassica plants are not the only plants that make seeds. Cherry trees make seeds. Where are they found?

There is one seed inside each cherry. And where does the cherry grow? It grows right where the cherry flower was.

Plants grow flowers. The flowers grow into **fruit**. Fruit have seeds inside. When the seeds grow into new plants, it is called **reproduction**.

Apple tree

Apple flowers

Apple seeds

Apple fruit

Have you ever seen tomato
flowers? Tomato flowers grow
into tomatoes. Tomatoes are
fruit. They have seeds.

Can you see the strawberry flowers?
Strawberry flowers grow into fruit.
Strawberries have seeds, too.

New plants grow from seeds. Seeds are found in fruit. Fruit grow out of flowers. Flowers and fruit are important in the life cycles of plants.

Thinking about Flowers and Seeds

1. Name one plant, and tell about its flowers.

2. Where are the seeds on a brassica plant?

3. Name two fruits you like to eat.

4. Name the parts of a plant that are important for its life cycle.

How Seeds Travel

How can we make sure plants have
the space they need? Get rid of weeds!
Weeds are unwanted plants.

How do weeds get into gardens?

Most weeds start as seeds. Seeds come
from flowers. First, the seeds get ripe.
Then, they are ready to travel!

Some seeds glide or spin in air. They might land far away. If they land on moist soil, they can grow.

Some seeds are carried by animals. These
seeds have little hooks. The hooks can hold
onto an animal's fur. The seeds go where
the animal goes.

Some seeds can even be carried by you!
They can stick to your sweater or shoes.
Some seeds will fall off. When they land
on moist soil, they can sprout and grow.

Birds and squirrels can move seeds, too. Birds eat berries and fly away. There are seeds inside the berries. The seeds pass through the birds. Now the seeds are in new places!

Squirrels eat seeds, too. They hide acorns to eat
during winter. Lost and forgotten acorns can grow
into oak trees. Seeds travel in many ways.

Now can you tell how weeds get into gardens?

Thinking about
How Seeds Travel

1. How do seeds travel in air?

2. How do seeds with hooks travel?

3. How do birds move seeds?

4. How do squirrels move seeds?

So Many Kinds, So Many Places

This amazing animal is an insect. Flies, ants, and crickets are all insects, too. There are so many kinds of insects. Insects are everywhere! Can you name some others?

No matter where you are, an insect is
probably near you. Insects are in the
air and in the water. Some creep in the
Arctic snow. Others scamper around in
the desert.

These ladybugs have gathered on a tree trunk. Some insects live on the tops of mountains. Other insects live in the rain forest. Insects are everywhere!

Insects might seem like pests to you. Some insects eat clothes, buildings, and crops. But insects are very important. Many different animals need them for food.

Insects are important for people, too. Bees move **pollen** from one flower to another, and that allows plants to grow fruit and seeds. Bees also make sweet honey.

People use thread from the cocoon of the silkworm to make clothing.

Next time you go outside, look for insects. They are everywhere!

Thinking about So Many Kinds, So Many Places

1. What are some ways insects are important to humans?

2. What are some ways insects are important to other animals?

Insect Shapes and Colors

Insects are different shapes and colors. The shape or color can help insects hide. An enemy might not see an insect that looks like its habitat. A hungry bird or lizard might think this insect is a leaf.

This praying mantis hides in the leaves, waiting to catch an insect to eat.

The walking sticks on this twig are very hard to see. Can you find them?

Look at the bright colors and design on this butterfly. Do you think it is hiding from its enemies?

Some insects are very easy to see. They are very colorful. They might have special markings.

Brightly colored insects often taste bad. They make other animals sick. Animals learn to stay away!

The spots on this beetle look like huge eyes.
A hungry animal might think the beetle is
a much bigger insect. The animal might be
scared away.

Insect Life Cycles

Insects might look different at each stage of their lives. Most insects go through four stages. The stages are egg, larva, **pupa**, and adult. The eggs of this insect were laid inside cells.

After a few days, a larva hatches from each egg. The tiny larva stays curled up inside the cell. The larva eats food made from pollen and honey. This food makes the larva grow.

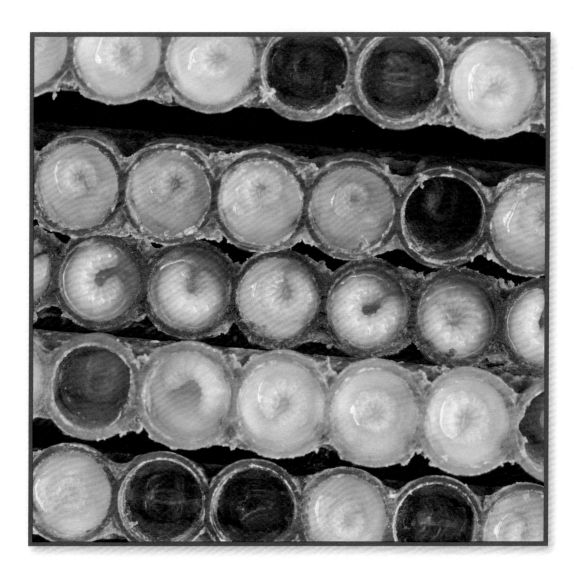

Then, the cell is covered with wax. Inside the cell, each larva turns into a pupa.

In the pupa stage, the insect goes through a big change. Soon, an adult crawls out of each cell. Do you know what insect this is?

It's a bee! After a short rest, the bee can go right to work. Young adult bees work in the hive. Older bees work outside the hive.

The larvae of different insects do not look the same.
These larvae will become insects you know well.
What will they look like as adults?

Moths and mosquitoes!

Some kinds of insects don't have larvae or pupae. When they hatch from eggs, they are called **nymphs**. Many nymphs look like their parents, but smaller.

Milkweed bugs go through four nymph stages.
In each new stage, they look more like an adult.
How many different nymph stages can you find?

Thinking about Insect Life Cycles

1. Tell about the life cycle of a bee.

2. Tell about the life cycle of a milkweed bug.

Life Goes Around

On a lucky day, you might see a ladybug.
A ladybug is red with black spots. This is
an adult ladybug. But have you ever seen
a baby ladybug?

Adult ladybugs mate. Then, the female lays
eggs. When an egg hatches, a larva comes out.
The black larva is a baby ladybug. But it doesn't
look like its parents. The larva eats and grows
for about 4 weeks.

Then, the larva **pupates**. Inside the pupa, the larva is changing. When the pupa opens, an adult ladybug comes out. It is red with black spots. Now it looks just like its parents.

The ladybug life cycle is like the life cycle of many other insects. It is like the life cycle of mealworms. It is like the life cycle of butterflies and moths. But it is different from the life cycle of some other animals.

Some animals hatch from eggs. Some animals are born alive. They all grow up to be adults. The adults mate and have babies called **offspring**.

Every animal goes around the life cycle. Cycle means to go around. The life cycle of a robin looks like this.

Robin eggs

Baby robins

Young robin

Adult robin

Trout lay eggs in streams. After 6 to 8 weeks, the eggs hatch. Tiny, fat babies swim out. You can see that they are fish. But they don't look like their parents yet.

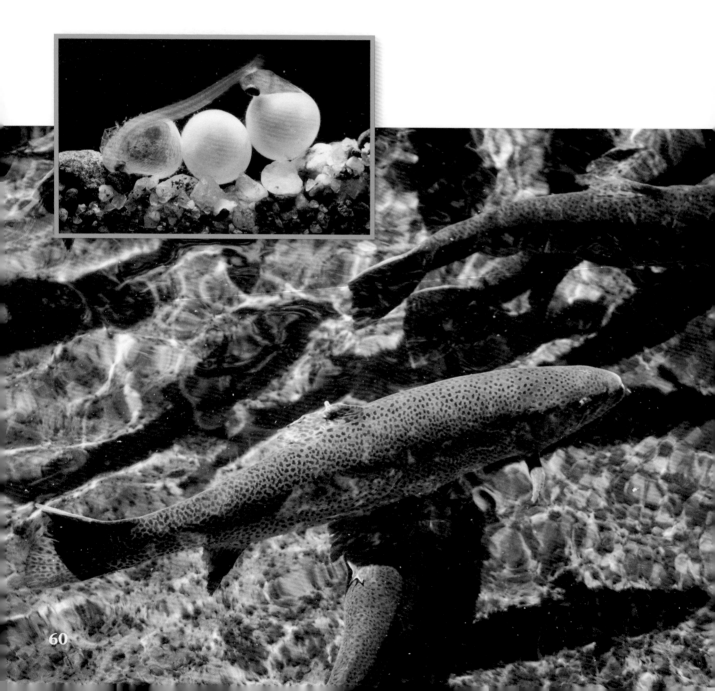

For the next year, they grow up little by little. In 2 years, they are adults. They look just like their parents. They mate and lay eggs in streams. Can you describe the trout life cycle?

Frogs lay eggs in water, too. When an egg
hatches, a tadpole swims out. It looks more
like a fish with a big head than a frog. It
doesn't look like its parents yet.

The tadpole eats and grows. In a few weeks, the tadpole starts to change. Its long, flat tail gets shorter. Its legs start to grow.

In a few more weeks, the tadpole has grown into a frog. It looks just like its parents. Can you describe the frog life cycle?

Ducks lay eggs in a nest in a **marsh**. The mother duck sits on the eggs to keep them warm. When they hatch, the babies are fluffy and yellow. The babies are called ducklings. You can see that they are ducks. But they don't look like their parents yet.

The ducklings eat and grow. In a few weeks, they get their brown feathers. In a few months, they are adults. They look just like their parents. In the next year, the adult ducks will mate. They will raise new families of ducklings. Can you describe the duck life cycle?

Mice do not lay eggs. Baby mice grow inside
the mother. The babies are born alive. Newborn
mice are pink, hairless, and blind. You can see
that they are mice. But they don't look like their
parents yet.

In a few days, the babies open their eyes. Their fur starts to grow. In a few months, they will be adults. They will be ready to continue the life cycle. They will have babies of their own. Can you describe the life cycle of mice?

Thinking about Life Goes Around

1. Does a ladybug larva look like its parents?

2. Tell about the life cycle of a ladybug.

3. Tell about the life cycle of a different animal.

4. Name five animals that hatch from eggs.

5. Name three animals that are born alive (not from eggs).

Glossary

adult a fully grown organism (16)

animal a living thing that is not a plant (5)

basic need something that is needed for plants and animals to survive. Plants and animals need air, water, food, space, shelter, and light. (4)

desert a dry place with little rain (7)

egg the first stage of a life cycle (15)

flower the part of a plant that grows into fruit (18)

fruit the part of a plant with seeds in it. Flowers grow into fruit, and fruit produce seeds in plant reproduction. (22)

grassland a place with a lot of grass and often no trees (3)

habitat the place or natural area where plants or animals live (3)

insect an animal that has six legs and three main body parts. They are the head, thorax, and abdomen. (5)

larva (plural **larvae**) a stage in the insect life cycle after hatching from eggs. Insect larvae look different from their parents and are often wormlike. (16)

life cycle the stages in the life of a plant or animal (18)

living alive (3)

marsh soft, wet land that is sometimes covered with water (64)

nutrient something that living things need to grow and stay healthy **(6)**

nymph a stage in the insect life cycle that has no larva or pupa. Nymphs look like their parents, but are smaller. **(52)**

offspring a new plant or animal produced by a parent **(59)**

plant a living thing that has roots, stems, and leaves. Plants make their own food. **(4)**

pollen a fine powder produced by flowers. Pollen is needed to produce fruit and seeds. **(39)**

pupa (plural **pupae**) a stage in the insect life cycle between the larva and adult stages **(46)**

pupate to change into a pupa **(57)**

reproduction the process of producing offspring **(22)**

seed the part of a plant found inside fruit. Seeds can grow into new plants. **(10)**

survive to stay alive **(11)**

thrive to grow fast and stay healthy **(17)**

tundra a place in the arctic or high on mountains **(12)**